DISCARD

DATE DUE			
C-B69			

581
DAM

Damon, Laura.

Wonders of plants
and flowers.

HACIENDA MAGNET SCHOOL
1290 KIMBERLY DRIVE

997322 01089 50978B 03093E

Wonders of Plants and Flowers

Written by Laura Damon

Illustrated by Yoshi Miyake

Troll Associates

Library of Congress Cataloging-in-Publication Data

Damon, Laura.
    Wonders of plants and flowers.

    Summary: Introduces the world of botany, explaining
simply the parts of plants, what they need to grow,
and how they propagate.
      1. Plants—Juvenile literature.   2. Flowers—
Juvenile literature.   [1. Plants.   2. Flowers.
3. Botany]   I. Miyake, Yoshi, ill.   II. Title.
QK49.D325   1990   581   89-5003
ISBN 0-8167-1761-3 (lib. bdg.)
ISBN 0-8167-1762-1 (pbk.)

It's usually easy to find plants—there are so many kinds, and they grow in many different places. Most of the earth makes a good home for plants.

Green forests, for instance, cover much of the land, both in warm, tropical climates and in cooler places.

High on windswept mountains grow strangely bent trees shaped by strong winds. And close to the ground, protected from the wind, grow brightly colored wild flowers.

Even in the icy tundra, the cold treeless land of the Arctic, flowers and mosses grow during the short, cool summer.

If you were to visit a sandy desert, where it is hot and there's little rain, you would still find plants. Sharp-thorned cactus plants dot the dry desert. And how does the cactus live without much water? Its fleshy stem stores water, which the plant uses a bit at a time.

Grasslands are another place where plants grow. As their name says, these large, open lands are covered by different kinds of grasses. Where the soil is rich, farmers use grasslands to grow much of the food we eat, like oats and wheat.

Can you think of another place where plants grow? The watery parts of the world are filled with plants. Seaweed and tiny microscopic plants make their home in the ocean. They make up much of the food that ocean animals eat. And in ponds, lakes, and streams, many other plants grow.

Everywhere we look in nature, we are almost certain to find plants. And all of them need the same three things to stay alive: soil that is rich enough to nourish them, temperatures that are warm at least part of the year, and water.

Scientists are not sure exactly how many kinds of plants there are, but they think there are at least 350,000 different ones.

Plants also come in many sizes. The largest are the giant sequoia trees of California. One of them is more than thirty feet wide and over two hundred ninety feet tall. Yet diatoms, tiny water plants, are so small that five hundred of them can fit into a drop of water.

There are two major groups of plants. The first group is
made up of plants that have special parts called roots, stems, and
leaves. These parts carry water and food through the plant.
Plants such as garden flowers, wild flowers, oak and pine trees,
ferns, bushes, shrubs, weeds, and many more belong to this
group.

Roots

Fungus

Algae

The plants of the second major group have no roots, stems, or leaves. Fungi, such as mushrooms, belong to this group. Water plants called algae are a part of this group, too. Seaweed is one type of algae you may have seen washed up on the shore. Other kinds of algae grow in ponds or on damp rocks or soil. Mosses also belong to this group. They often form a soft, low carpet of green upon the shady, damp rocks of a forest.

Among the plants that have roots, stems, and leaves are the flowering plants. These plants have flowers that create seeds. From the seeds, new plants grow. The flowering plants form a very big group—there are more than 250,000 kinds.

The seeds of most flowering plants are protected by a small shell or case. But there are certain plants with seeds that are not in such shells. These plants are the conifers, which grow their seeds in cones. Evergreens, such as pine and spruce trees, are conifers.

Some plants do not have flowers to make seeds. Ferns, with their delicate fanlike leaves, are flowerless. Instead of seeds, these plants have special parts, called spores, from which new plants grow.

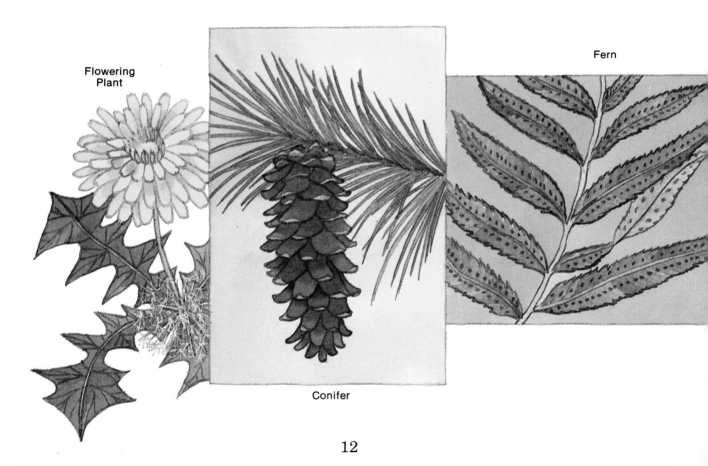

Flowering Plant

Conifer

Fern

Plants surround us with their beauty. They also give us many useful things like medicine, clothing, and lumber. But plants play an even more important role in our lives. Did you know that without plants, there would be no life on earth?

Plants are part of something scientists call the *cycle of nature*. A *cycle* is a sort of "circle." It happens over and over again, the way a wheel spins around. The cycle of nature has several steps that link together all living things—people, plants, and animals.

13   HACIENDA MAGNET SCHOOL

The cycle begins when plants absorb sunlight. The sun's energy runs the whole cycle. Plants use this energy in a special way. When they need nourishment, green plants use the sun's energy to help make their own food! This is something animals and people cannot do.

To make food, plants use energy from the sun. They also use a gas in the air called carbon dioxide, and water and minerals from the soil. This special process is called *photosynthesis*, a word that means "putting together with light."

During photosynthesis, plants give off oxygen into the air. This is important because people and animals need oxygen to breathe to stay alive. Then, when they breathe out, people and animals give off carbon dioxide and put it back into the air.

This is how the cycle starts again. We give plants carbon dioxide for photosynthesis, while plants give us oxygen for breathing.

In the cycle of nature, plants also give people and animals food to eat. In turn, people eat some of the animals that have eaten plants. So, in a way, all our food comes from plants.

Then, when animals and plants die and decay, or break down, many minerals are put back into the soil. The soil is rich once again. It is ready for new plants to grow so that the cycle of nature can go on.

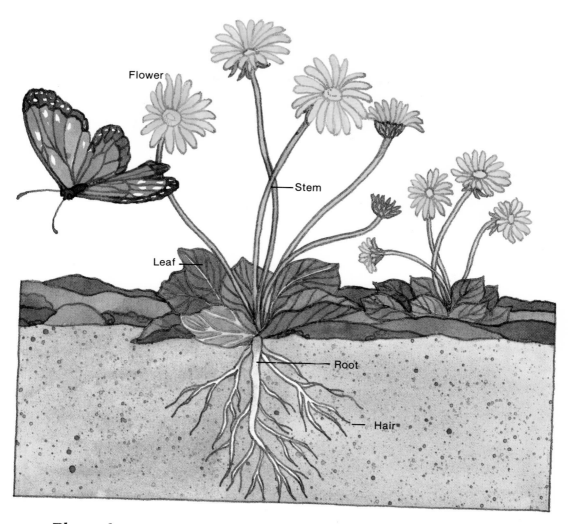

Flower

Stem

Leaf

Root

Hair

Plants have special parts that help them live and grow. The flowering plants, the most common group, all have the same four parts. These are roots, stems, leaves, and flowers.

Each of these parts plays a special role in the life of a plant. *Roots* anchor the plant. They usually grow underground to hold the plant in place. Special tiny hairs, called root hairs, grow on each root. Their job is to take up water and bits of mineral food from the soil.

From the roots, water and minerals travel to the plant's *stem*. The stem holds up the plant to the sunlight and gives the plant its shape.

Trunks, branches, and twigs are all stems. Some stems, such as the stem of a dandelion plant, are short and soft. Other stems are large and hard, such as the trunk of a big maple tree.

Inside the hard branches and trunks of trees and shrubs, there is a thin layer of material called *cambium*. The cambium is where new cells are made. Each year the cambium layer grows a new ring inside the tree trunk. So if you should see the stump of a tree, count the cambium rings. They will tell you how many years old the tree was when it was cut down.

On their journey through the stem, life-giving water and minerals pass to the *leaves* of a plant, where photosynthesis takes place. The leaves have a green material called *chlorophyll*, which gathers energy from sunlight to help the plant make food.

A leaf starts on a stem as a bud. Leaves grow in a great many sizes and shapes. Just look around you and compare the many different leaves you see. Most leaves are flat and broad. They have a pattern of tiny tubes, or veins, on them. Water passes through these veins. Many tiny openings cover the top and bottom of each leaf. Misty water vapor, plus certain gases, pass in and out of the leaf through these holes.

The fourth main part of a plant is its *flower*. In spring and summer, beautiful blossoms can be seen.

Some flowers, like tulips, have brightly colored, large flowers. Other flowers are small and much plainer. In the spring, look closely at the first yellowish-green growths on a maple tree. You will see they are really a cluster of tiny flowers.

The largest flower of all is the rafflesia, which grows in the hot, wet jungle. This giant bloom can grow up to four feet wide and may weigh more than fifteen pounds.

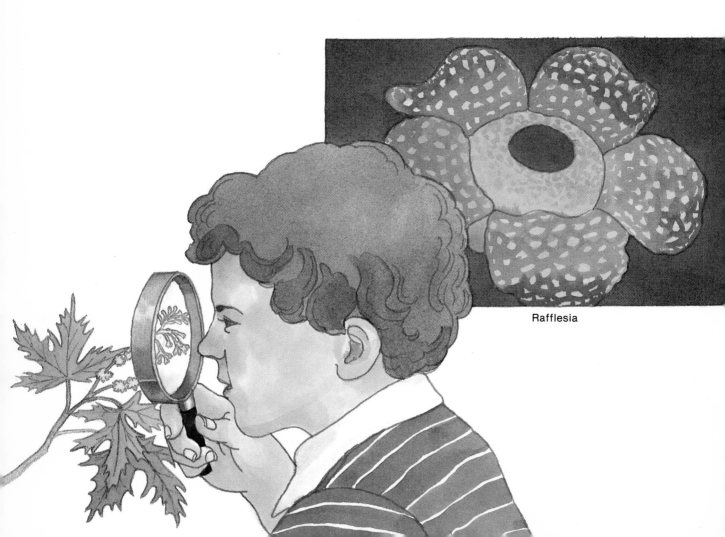

Rafflesia

Flowers grow from buds on a stem. Inside each flower are the parts needed to make seeds. From the seeds, new plants will grow.

Most flowers have four parts. The *calyx* has leaflike parts, which protect the flower bud growing inside the calyx. The flower's most colorful part is usually the *corolla*, or petals. The *stamens* are the male part of the plant; they have a material called pollen in them. *Pistils* are the female part of the plant. At the base of the pistil is a case where tiny eggs are found.

For seeds to form, pollen from the stamen must be joined with an egg in the pistil. This joining is called *pollination.*

How does pollination take place? Many flowers have helpers. The wind is one of them. It may blow pollen from one flower to the pistil of another flower. Other helpers are bees, birds, bats, and butterflies.

The odors and bright colors of some flowers attract these helpers. Honeybees especially like blue and purple flowers. And these bees can tell the difference between more than thirty kinds of flowers by the way each blossom smells. As they visit each flower, bees, bats, tiny hummingbirds, and butterflies sip the sweet liquid, called nectar, inside the blossom.

While collecting nectar, sticky pollen grains may cling to the legs, heads, or bodies of these creatures. Then the bee or animal flies off to another flower. Some of the pollen from the first flower may rub off the insect or animal onto the pistil of the next flower, pollinating it. If this happens, seeds will grow.

After the blossom withers and falls away, the seeds are
protected by a fruit that grows around them. Can you think of
some of the fruits you like to eat that have a pit or seeds inside
them? Plums, berries, apples, tomatoes, corn, and cucumbers are
just a few of them.

Inside each seed, a tiny new plant is waiting to grow. Seeds have many ways of traveling in order to find a good growing place.

The coconut is a large seed that can float. It falls from the tree near the ocean shore. Then it floats far away to another shore, where it takes root and grows.

Some seeds are specially designed to be carried by the wind. Dandelion seeds, for example, are light and fluffy so the wind can carry them great distances.

Squirrels help oak trees to grow by burying acorns. They dig up and eat some of the acorns, but not all of them. In spring, the forgotten acorns take root and sprout as young trees.

People also spread many seeds all over the world by planting crops—plants are a big part of our diets!

Plants do not seem to be moving, but they are. Some plants close their flowers or leaves at night, opening them once again in the morning light.

Plants also move toward sunlight. If a plant is placed in a sunny window, it will turn its leaves and branches toward the light. If the plant is turned in the opposite direction, all the leaves and branches will soon turn around, facing the sun once again.

The roots of a plant also move in a certain way: they always grow downward. Even when seeds are planted upside down, their sprouting roots will turn in a downward direction.

Some plants live in places where they cannot get the mineral food they need. These plants have a special way of moving to get mineral nourishment—they capture and eat insects!

A plant called Venus' flytrap has leaves that can open and shut like a hinged door. If a fly lands on one half of the leaf, the two halves close tightly, trapping the insect. Juices in the leaves break down and digest the insect, so the plant can use it for food.

Some plants are harmful not just to insects, but to people, as well. There are plants that cause allergies, and weeds that choke important food crops.

But most plants give us beauty and useful products. Even the plants that lived long ago are helpful to us today. These ancient plants have been changed over millions of years into some of the fuels we use, such as coal, oil, and gas.

No matter where you live, you can be a botanist. A botanist is someone who studies plants. Look at the plants around you. You'll very likely make some interesting and beautiful discoveries. For each plant is fascinating in its own way. And each one plays an important part in the story of our planet.